STARTING OUT

baby BEARDED DRAGONS

LAURA K. MURRAY

CREATIVE EDUCATION • CREATIVE PAPERBACKS

CONT

ENTS

I AM A HATCHLING.

I am a baby Bearded Dragon.

Look at
my bumpy
scales!

I hatched from an egg. My mother laid eggs in a burrow.

It took 85 days for me to hatch. I was only 3 inches (7.6 centimeters) long!

Bugs and plants make up my meals. Crunch! My large tank gives me room to move.

I warm up my body by basking.

My colors brighten.

I puff up my beard to look bigger. I shed my old skin as I grow.

Time for some new skin!

SPEAK AND LISTEN

SSSSSSS!

Can you speak like a hatchling? Bearded dragons may hiss when they are angry. They also bob their heads and wave their arms.

Listen to these sounds:

Now it is your turn!

HATCHLING WORDS

Bask: lie on a warm surface to raise body temperatures

Burrow:: a hole or tunnel

hatch: break out of an egg to be born

shed: get rid of old skin and make way for new skin

READING CORNER

Hansen, Grace. *Bearded Dragons in the Wild!* Minneapolis: Pop!, 2024.

Kingsley, Imogen. *Bearded Dragons*. Mankato, Minn.: Amicus Ink, 2019.

Lynch, Seth. *Bearded Dragons*. New York: Gareth Stevens Publishing, 2024.

INDEX

PUBLISHED BY CREATIVE EDUCATION AND CREATIVE PAPERBACKS
P.O. Box 227, Mankato, Minnesota 56002
Creative Education and Creative Paperbacks are imprints of The Creative Company
www.thecreativecompany.us

LIBRARY OF CONGRESS CATALOGING-IN-PUBLICATION DATA
Names: Murray, Laura K., 1989- author
Title: Baby bearded dragons / by Laura K. Murray.
Description: Mankato, Minnesota : Creative Education and Creative Paperbacks, [2026] | Series: Starting out | Includes bibliographical references and index. | Audience: Ages 4-7 | Audience: Grades K-1 | Summary: "Introduce beginning readers to the world of baby bearded dragons with this life science starter. Includes photos, a labeled animal diagram, "Speak and Listen" section, glossary, and further resources"-- Provided by publisher.
Identifiers: LCCN 2025015253 (print) | LCCN 2025015254 (ebook) | ISBN 9798895810309 library binding | ISBN 9781682779835 paperback | ISBN 9798895811566 ebook
Subjects: LCSH: Bearded dragons (Reptiles)--Infancy--Juvenile literature | Bearded dragons (Reptiles)--Juvenile literature
Classification: LCC QL666.L223 M86 2026 (print) | LCC QL666.L223 (ebook) DDC 597.95/5--dc23/eng/20250618
LC record available at https://lccn.loc.gov/2025015253
LC ebook record available at https://lccn.loc.gov/2025015254

DESIGN AND PRODUCTION
Design by Rhea Magaro
Art direction by Graham Morgan

PHOTOGRAPHS BY
Dreamstime/Alex Bramwell, 5, Isselee, 8, Jacqui Jones, 11, Jesue92, 4, Marie Elena Sager, 12, Marijus Auruskevicius, 2; Getty Images/Daniel Farkaš, 14, Pablo Reinsch / 500px, 6, Witold Steblik / 500px, 10; Shutterstock/Robert Eastman, cover; Wikimedia Commons/HTO, 7, Max Tibby, 9

Printed in the United States

Archaeology of Domestic Architecture and the Human Use of Space

For Girish and Rose, as always